P9-CKD-587

ROBOTS

Consultant
Dr. Howard L. Harrison
Professor of Mechanical Engineering
University of Wisconsin — Madison

Library of Congress Number: 80-11681

1 2 3 4 5 6 7 8 9 0 84 83 82 81

Printed and bound in the United States of America.

Library of Congress Cataloging in Publication Data

Kleiner, Art.
Robots.

Includes index.
SUMMARY: Describes how a robot operates and discusses the development, uses, and future of these machines.
1. Automata — Juvenile literature. [1. Robots]
I. Scott, Jerry, 1941- II. Title.
TJ211.K56 629.8'92 80-11681
ISBN 0-8172-1401-1

ROBOTS

By Art Kleiner
Illustrated by Jerry Scott

CONTENTS

RAINTREE PUBLISHERS
Milwaukee • Toronto • Melbourne • London

WHAT IS A ROBOT?

The best-known robots today are R2-D2 and C-3PO of the movie *Star Wars*. Like most robots in movies or television, they walk, talk, and think. They can translate languages or uncode complicated charts. They can sense what the people around them are feeling. They can fly rocket ships, march across the desert, and even make jokes.

The robots in movies like *Star Wars* are not real. Even R2-D2 was played by a small man in a robot suit. But there are already real robots which "think," move by themselves, and work for us. This book will tell you about what they are doing and how they work.

People do not agree on exactly what a robot is. Some scientists say a robot must have some kind of brain and be able to think for itself. Other scientists say that every machine, even a refrigerator or a bicycle, is a kind of robot.

Most scientists who work with robots say this: A robot is a machine body, controlled by a computer brain, which operates by itself and can be taught to do different kinds of work.

A machine is a tool which helps people do things they can't do by themselves. For example, a wheel is a machine which rolls. A wagon is a machine made of four wheels which carries things that are too heavy for people. A bicycle is a more complicated machine; it is made up of about 20 simple machines put together. A robot's body is even more complicated than a bicycle; it is made of hundreds of simple machines.

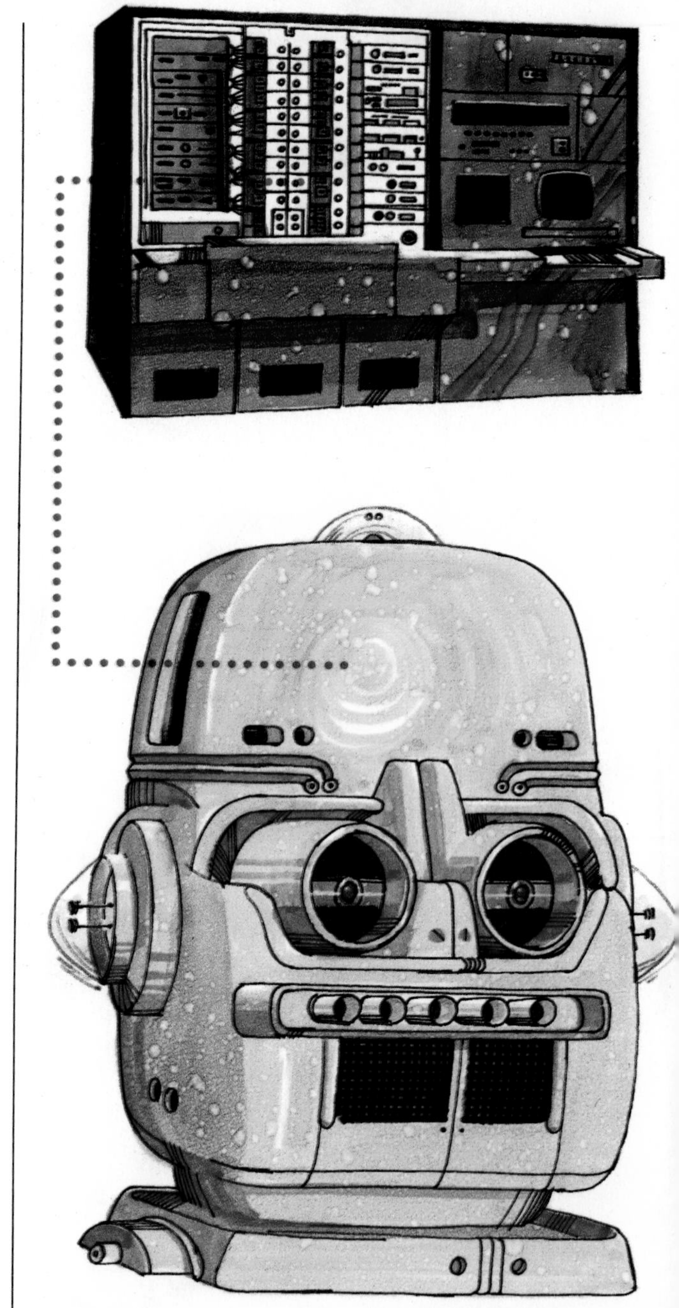

A machine needs some sort of brain to tell it what to do, the way a bicycle needs a person to ride it. If you hooked up a computer to your bicycle and the computer told it when to stop, go, and turn, then your bicycle would be a robot. Since most computers don't know how to balance, your bicycle would probably fall over.

A computer is an electronic machine which can do complicated arithmetic very quickly. It can translate ideas and commands into numbers, store the numbers away, bring them out when they are needed, and translate them back into ideas. It can store every kind of movement a machine can do.

Many robot movements are based on the movements of the human arm and hand. Robots can turn, lift things, put things down, and pull things. They can also hold tools: for instance, they can use a welding gun to join pieces of metal.

THE HISTORY OF ROBOTS

Modern robots are only about 30 years old, but machines like robots existed 1,000 years ago. In the Middle Ages, the clock makers of France and Germany began to build what they called automatons. These were small statues of people or animals which moved when the clock struck a certain hour or when a person turned them on. The cuckoo in a cuckoo clock is one kind of automaton.

By the 1800s the automatons were very complicated. Some automatons played musical instruments, drew pictures, or wrote words on a piece of paper. They were beautiful and lots of fun to watch, but each one could only do one type of movement. People hoped that one day machines would be built which could think for themselves and do our work for us.

The idea of smart machines doing our work also came from science-fiction stories. These are stories about things which might be possible in the future, but are not common in our lives today. The word robot itself was invented by a science-fiction writer, Karel Čapek, in the country Czechoslovakia in 1922. The word "robot" comes from the Czechoslovakian word for "work."

The robots in science-fiction stories and movies often looked like people. Some robots became so smart they got angry at the humans who made them work. Some got violent. But most science-fiction robots were simply machines, built to serve the humans around them. They were smart and could do many things, but they couldn't fight against the humans they worked for, any more than your bicycle could ride away all by itself with you on it.

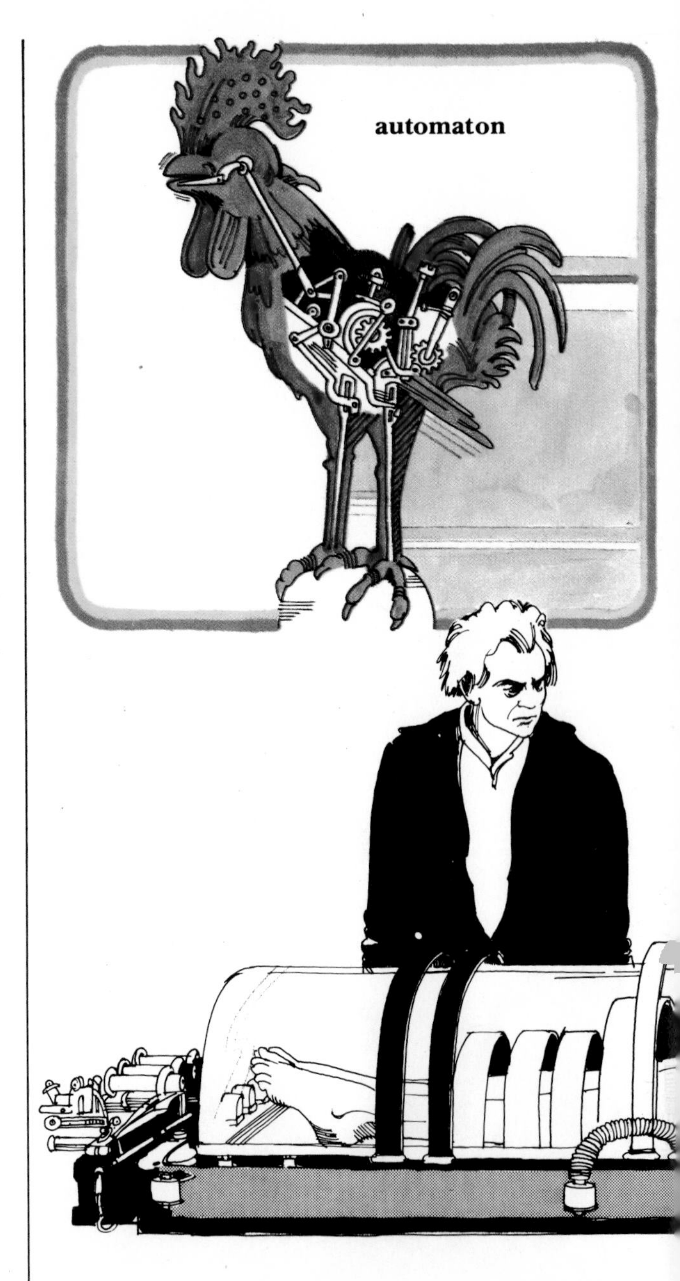

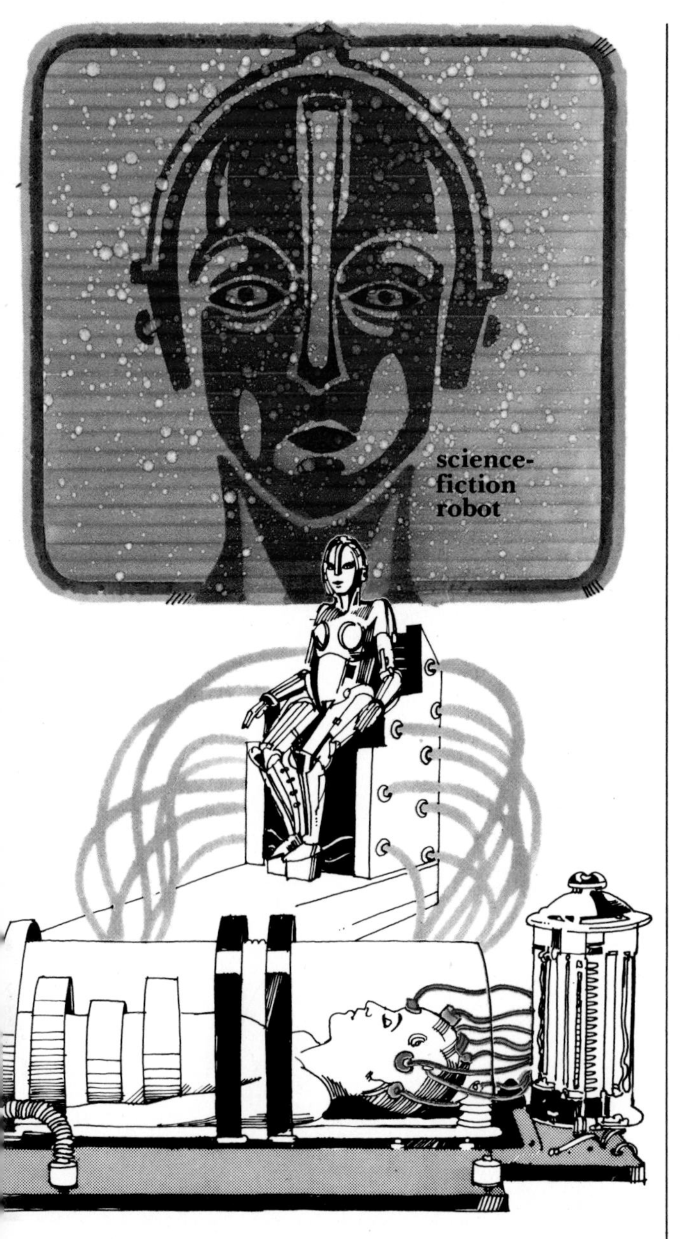

People began to make computers in the 1940s, and robots followed soon after. The first computers were as big as large rooms, too large to use with robots. But by 1960, some computers were small enough to become part of new machines. Scientists called these new machines "robots" after the smart machines in science-fiction. But real robots don't usually look anything like people. That would be too complicated to build. Some robots look like people, but they are not usually able to do much work.

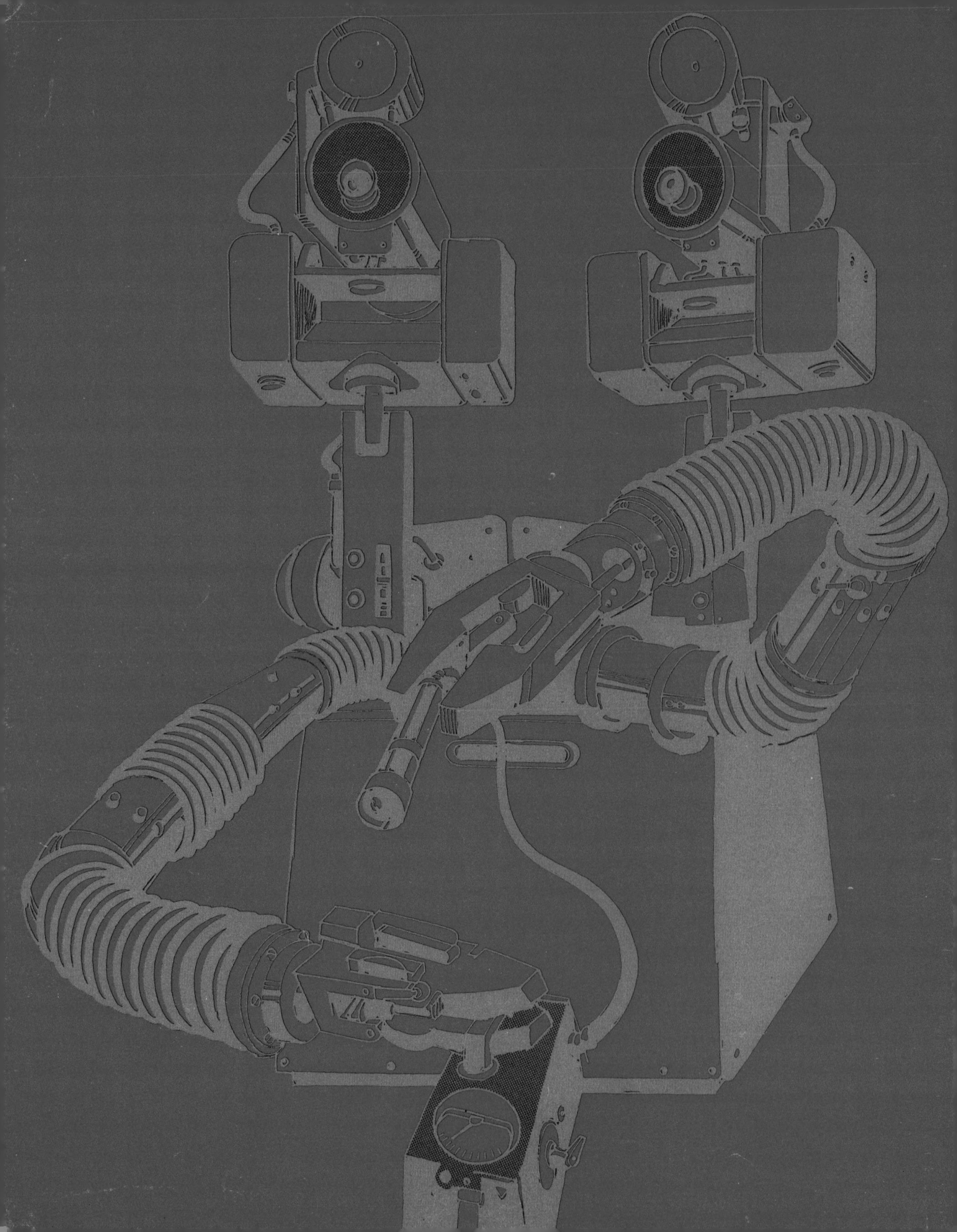

KINDS OF ROBOTS

There are some interesting kinds of robots in the world today. Some robotlike machines have no computers in them at all. They have bodies like robots, but they are operated by people. People control them the way a person controls a puppet on a string.

A machine called the Mobot has been used since World War II to work with chemicals that are too dangerous for people to touch. The Mobot has big metal arms and metal fingers that hold things the same way humans do. A human sits in another room and works the controls, and the Mobot moves as the person tells it to.

Some remote-control robots are attached to wheelchairs for people with no legs and arms of their own. Others are used to fight fires or dive under water. But they are not real robots, because every move is directed by a person. They don't do anything by themselves.

Many robots look like the Unimate® robot, made by a company called Unimation® Inc. These robots with small computers in them are the kind used in most factories today. They have one arm with a part like a hand. The body can turn in many directions, but cannot move from one spot.

Unimate robots and others like them work by themselves, doing the same set of movements over and over again. Many of them are used in assembly lines, to put cars and other machines together. They punch holes in metal or sort small metal parts, and they can use welding guns that fill the air around them with sparks.

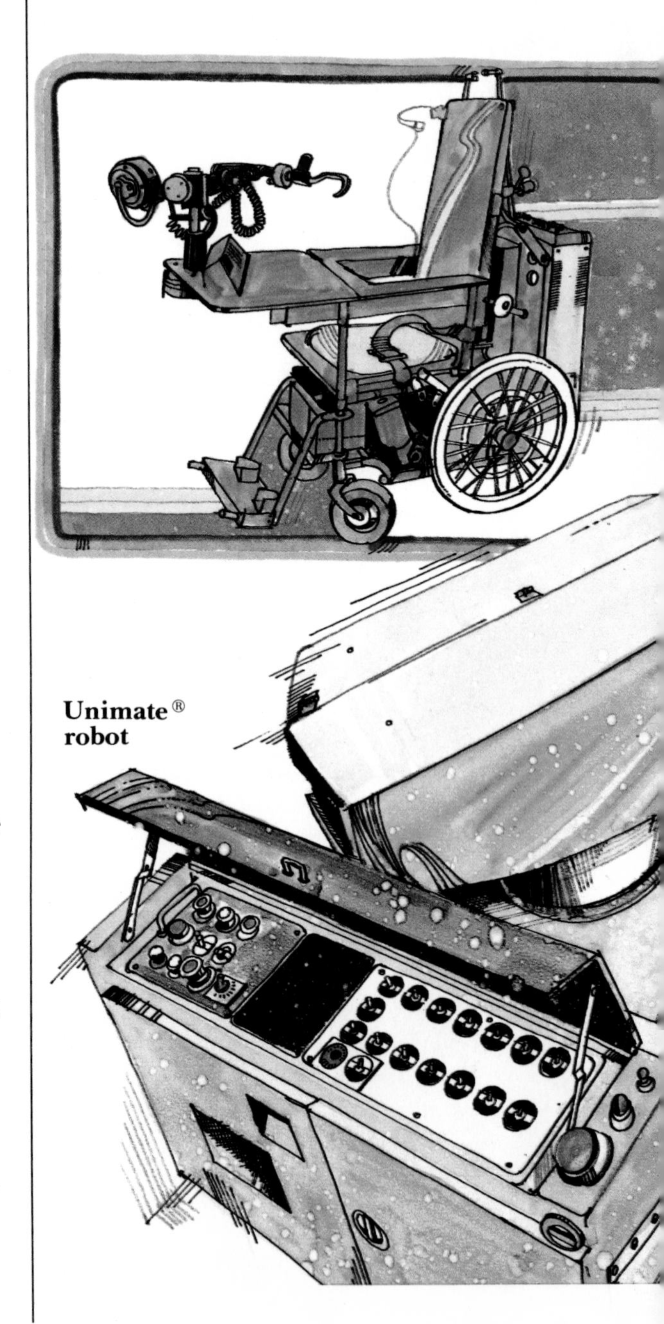

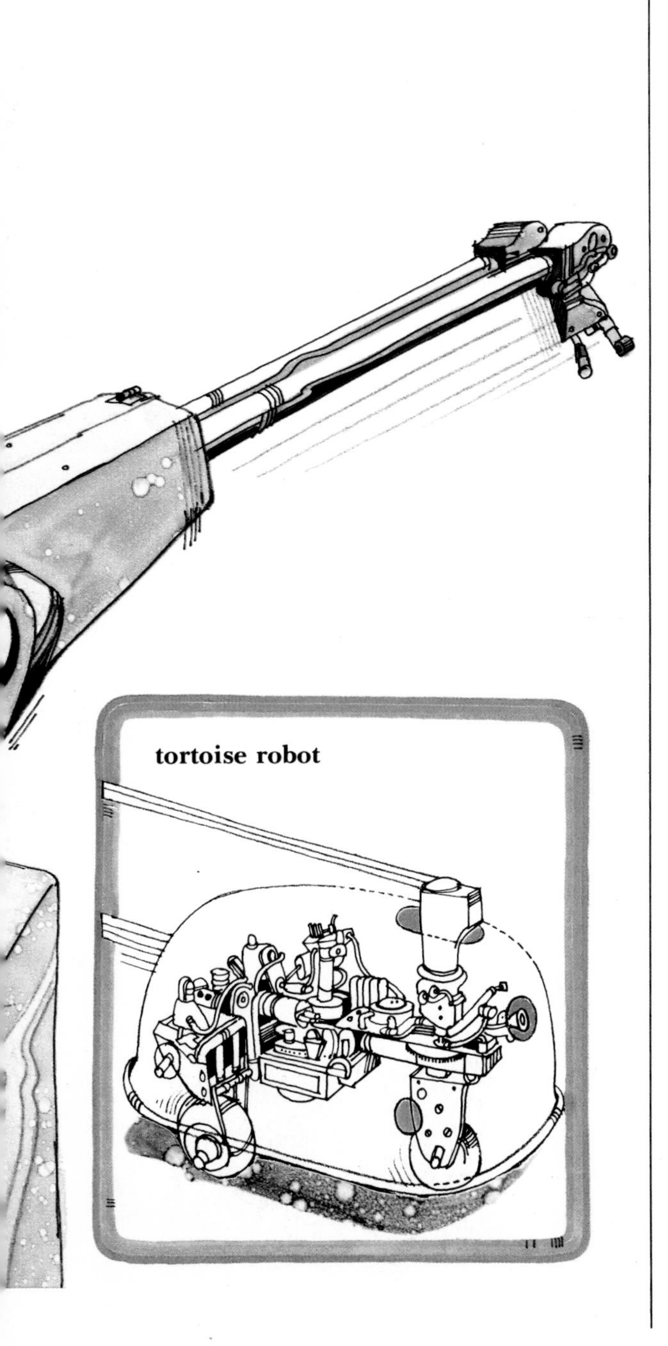

Scientists have tried to build robots which can think for themselves. To do this, robots need to be able to see, hear, or feel what is around them. Some robots do that. They are like little animals. The tortoise is a small machine on wheels. It travels to wherever it senses light. It will follow a beam of light around the room.

The other robot looks like a round box on wheels. It can tell where electric outlets are in the wall. When it finds one it moves up to it, plugs itself in, and recharges its batteries.

They led to the invention of a mail robot. Some companies use it in their offices to bring mail from one desk to another. The mail robot follows a magnetic path in the floor, and goes from room to room. It stops along the way whenever a built-in signal in the floor tells it to stop.

After a minute, it will contine on its way to the next desk. If anyone stands in its path, the robot senses them and pauses until they go away. If you took it off its magnetic path, however, it wouldn't know where to go. It would stop working.

Robots that can think are used in places where people can't go by themselves. Some of the most exciting robots are the ones used in space travel. When rockets were sent to land on Mars they did not carry people, but there were landers aboard, called Viking 1 and 2, that worked like robots. They used radio waves between Mars and Earth to get instructions for their computer and to send us pictures taken with their television cameras, or measurements taken with other instruments. They used mechanical arms to pick up soil samples. The arms took them back to the rocket ship to analyze them and send us the information about them. They also told us what air and gravity

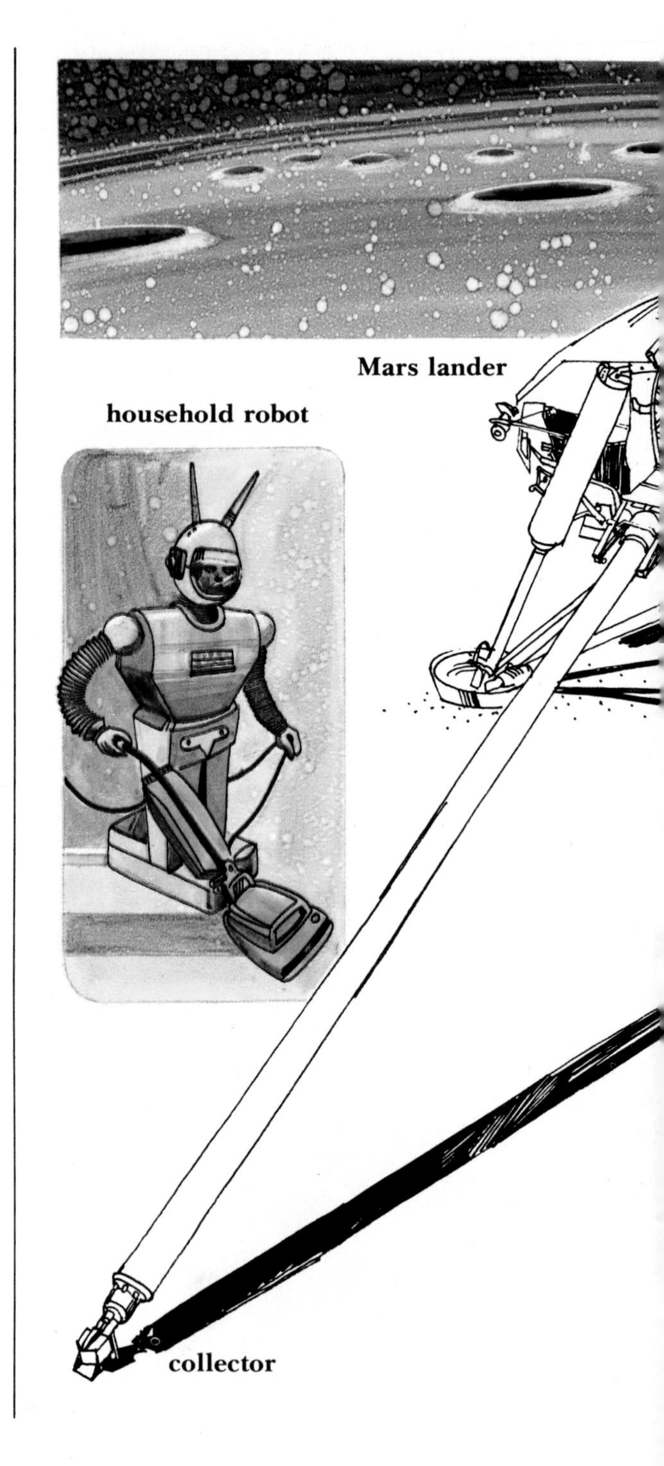

chess-playing robot

were like on Mars. Other robots may soon do the same thing on other planets or at the bottom of the ocean.

Some people say they are building robots which will serve us in our homes. They will clean the house, answer the door, and serve the food. Many inventors, including teenagers, have built working household robots and shown them at science fairs. But no one has yet made one that can be built cheap enough for people to afford.

However, a man named John Gallaher makes and sells kits of parts that can be put together into simple robots. This is his chess-playing robot. It must be connected to a computer that has been programmed with the rules and methods of the game of chess. When the computer decides on a move, the robot picks up the right piece and moves it on the board.

Entertaining robots, known as "Audio-Animatronics*" figures, perform daily in the shows and attractions of Disneyland and Walt Disney World. The figures range from realistic humans to animals, birds, and even animated flowers.

Disney "Audio-Animatronics" figures, such as the lifelike Abraham Lincoln, or those found in the "Pirates of the Caribbean" adventure, begin as clay sculptures. Plaster casts of these full-size figures are used to make the body forms which hold the internal mechanics.

Movements for the figures are computer programmed and recorded on tape along with spoken words, music, sound, and lighting effects. When the tape "plays back" the recorded information, the whole show is put together into a performance which can be repeated over and over again.

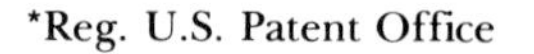

*Reg. U.S. Patent Office

Walt Disney pirate

Artoo-Detoo

See-Threepio

The golden, humanlike, talking robot, See-Threepio, was played by a man inside a fiberglass and metal costume, molded to fit his body perfectly. It was a very heavy costume to wear, and it was very hot during filming. Light pieces of the costume were placed on a dummy when it was not necessary for the actor to wear the whole costume.

The smaller robot, Artoo-Detoo, was a harder problem. The metal costume was moved by a midget who was inside for much of the film's shooting. However, Artoo's body could only be moved a few feet at a time, because the costume was heavy and difficult to work. Therefore, other copies were made for different purposes. There was a remote-controlled R2 unit that had blinking lights, as well as a hollow model carried by the Jawas, and a fresher, cleaner version that was used in the final scene of the movie.

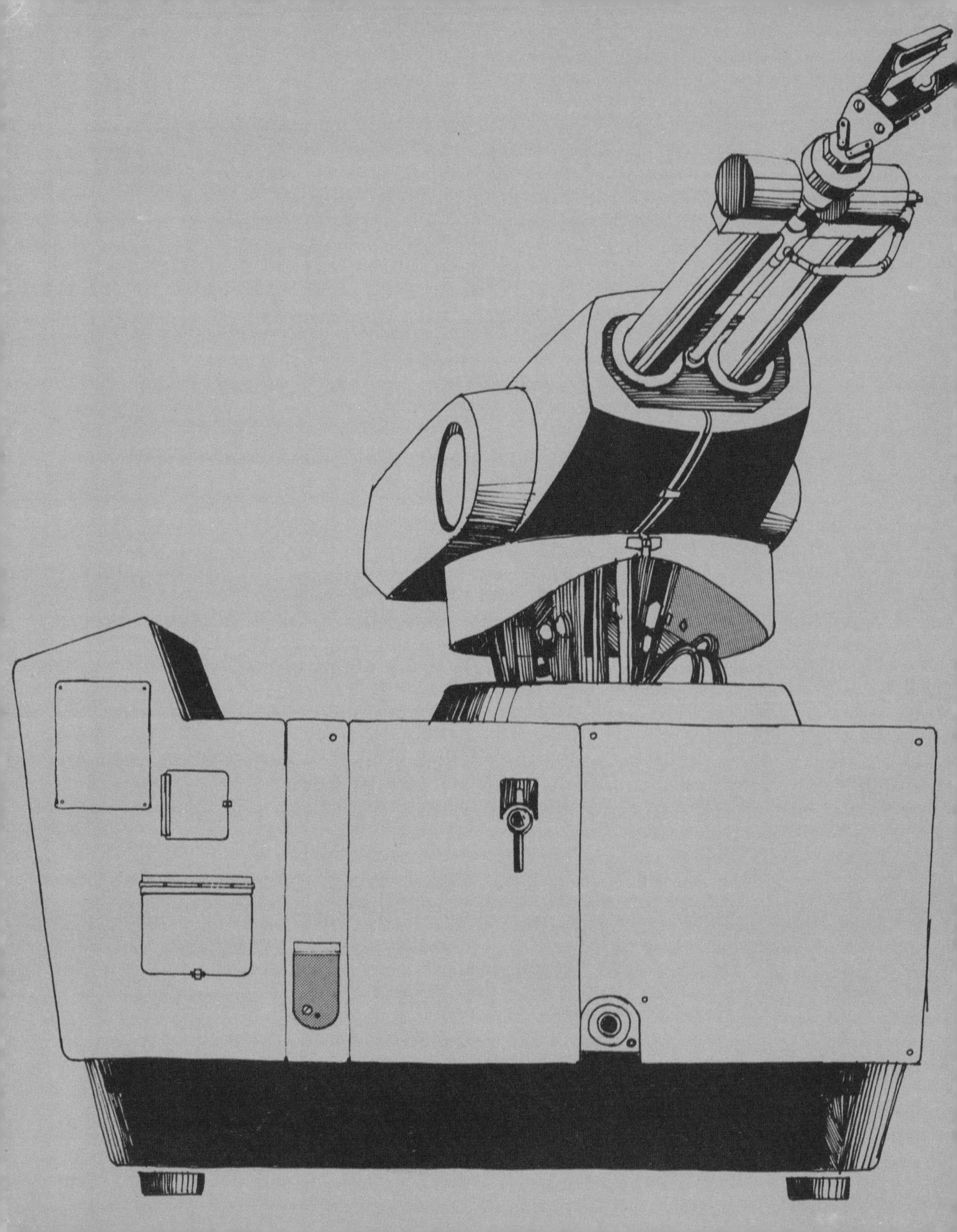

INSIDE A ROBOT

How would you build a robot? You would first build a computer brain, so it could think. Then you would need a machine body, so it could do things. You might want to make sure it could move around, and you would try to make it see, hear, or feel what is around it.

This Unimate robot is typical of most robots in factories. It is about 4.5 feet (1.4 meters) high, 5 feet (1.5 meters) long, and 4 feet (1.2 meters) wide. It is as tall as you are but much wider. It weighs about 3500 pounds (1588 kilograms). It is shaped like a big square box with a crane on top. A steel rod pulls in or pushes out from the crane. At the end of the rod is a robot hand, which can grip, hold, or push things. The crane can move the robot's arm up, down, or sideways.

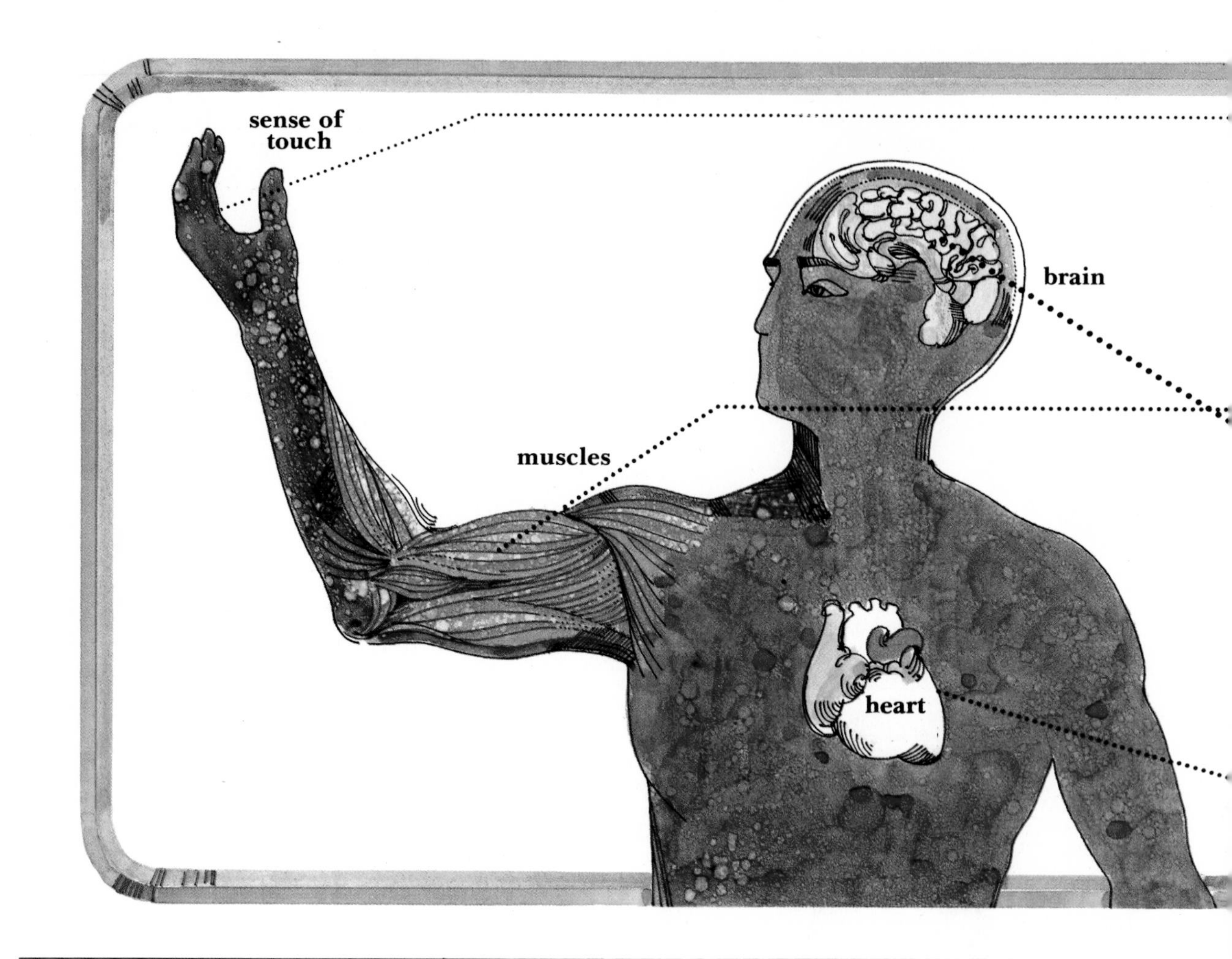

Each part of a robot is like a part of a human being. The computer is like a brain. The machine is like the body. Inside the machine, the motor and pump which give it power are like the heart. The cylinders which move the parts of the robot are like the muscles. The metal frame of the robot is like the bones. The wires that tell the computer how far the arm has moved are like nerves. Like our eyes, ears, and sense of touch, the robot has sensors. Sensors are the different machines which pick up information about what is near them.

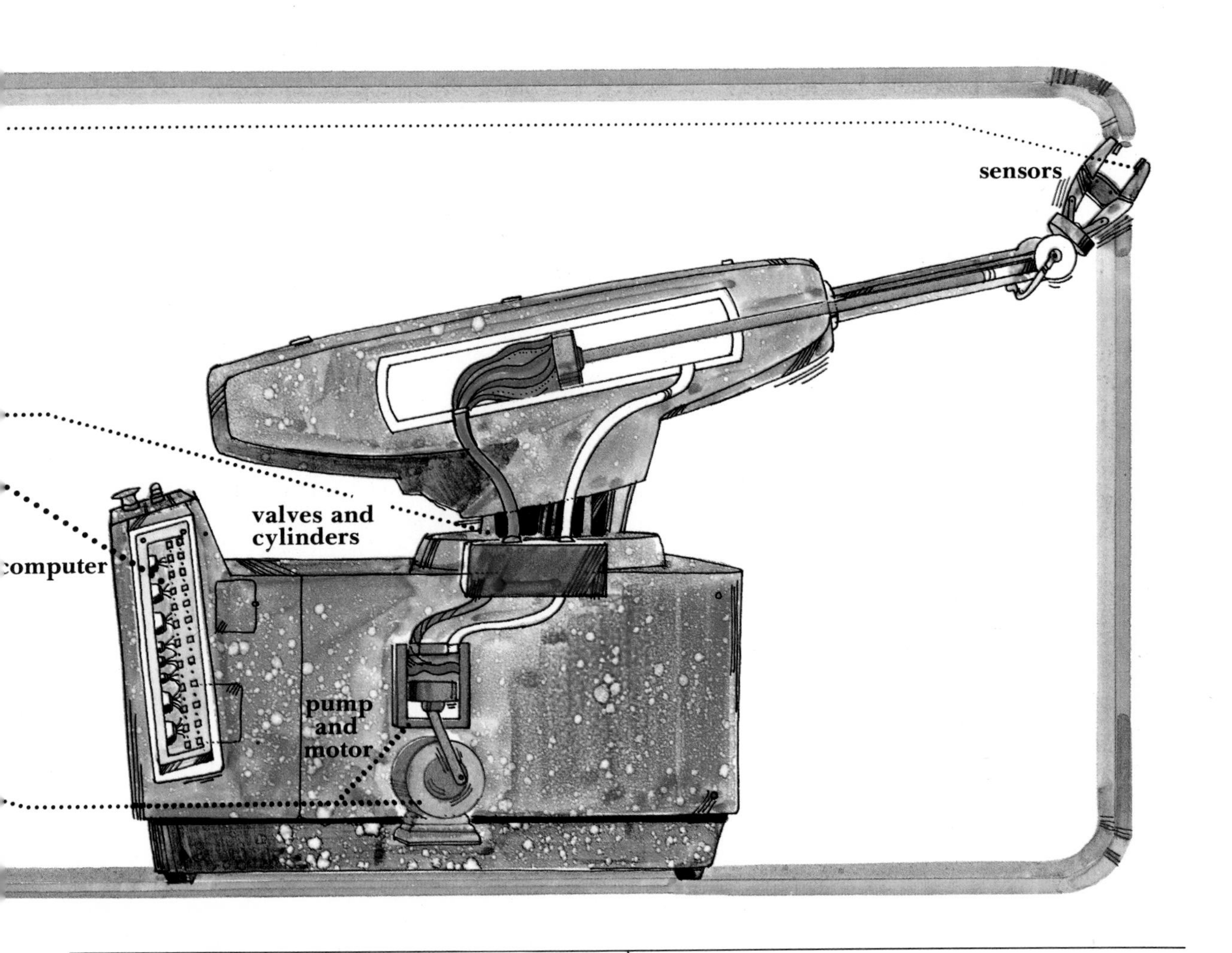

Electricity is the power which makes the robot move. Electricity passes through the computer to the motor, which pushes or pulls the parts of a robot. The motor is a machine inside the box at the bottom of the robot. When electricity passes through the motor, it makes parts in the motor spin around.

The spinning parts of the motor are connected to all the parts of the robot which have to move.

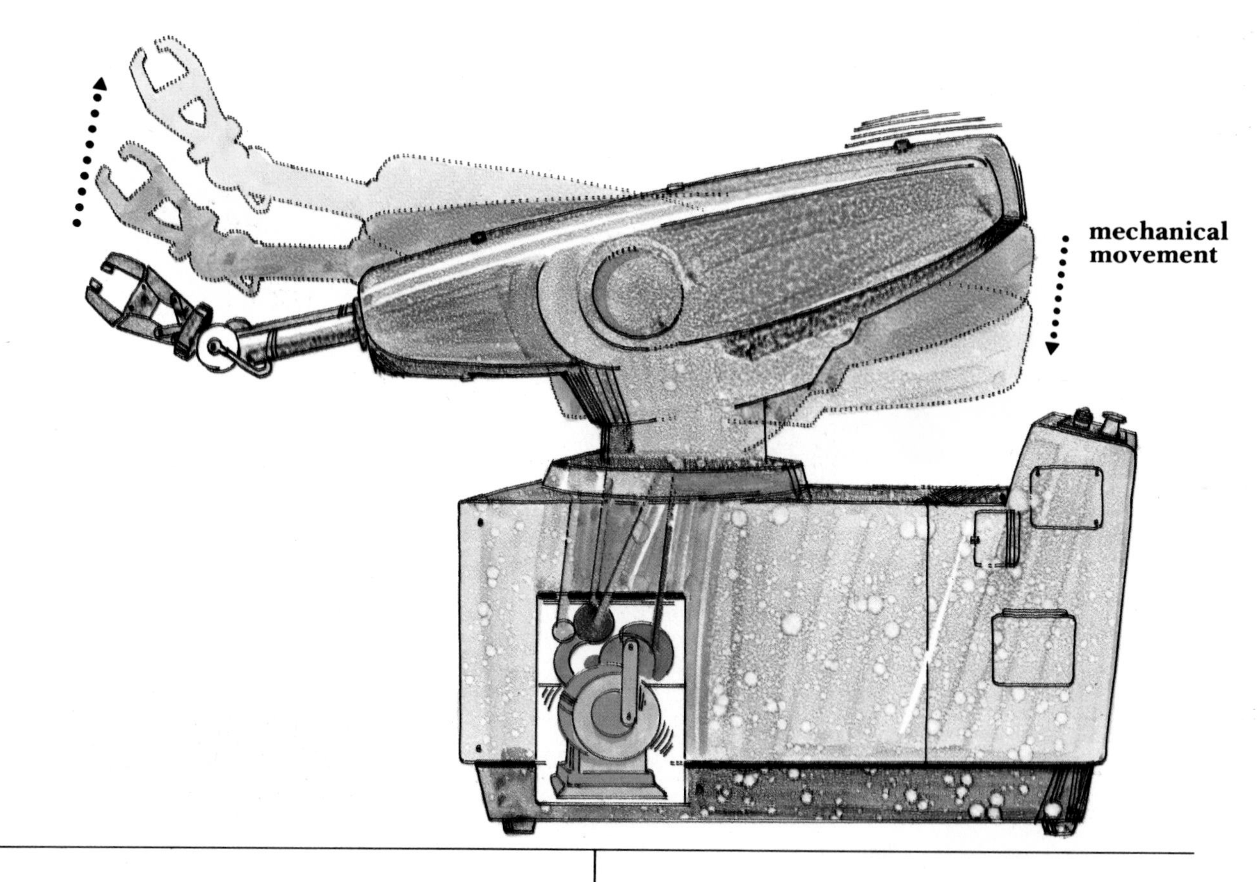

The motor can make the parts of the body move in three ways. One way is *mechanical*. The spinning parts in the motor are connected to other moving parts, such as gears or cables. Gears are wheels with parts like teeth, which catch in the teeth of other gears and move them. Cables are wires between other wheels. When one wheel moves, so does the other one.

People who make machines with gears and cables use different sizes of wheels and gears to get them to move at different speeds.

The old clockwork automatons moved all their parts mechanically.

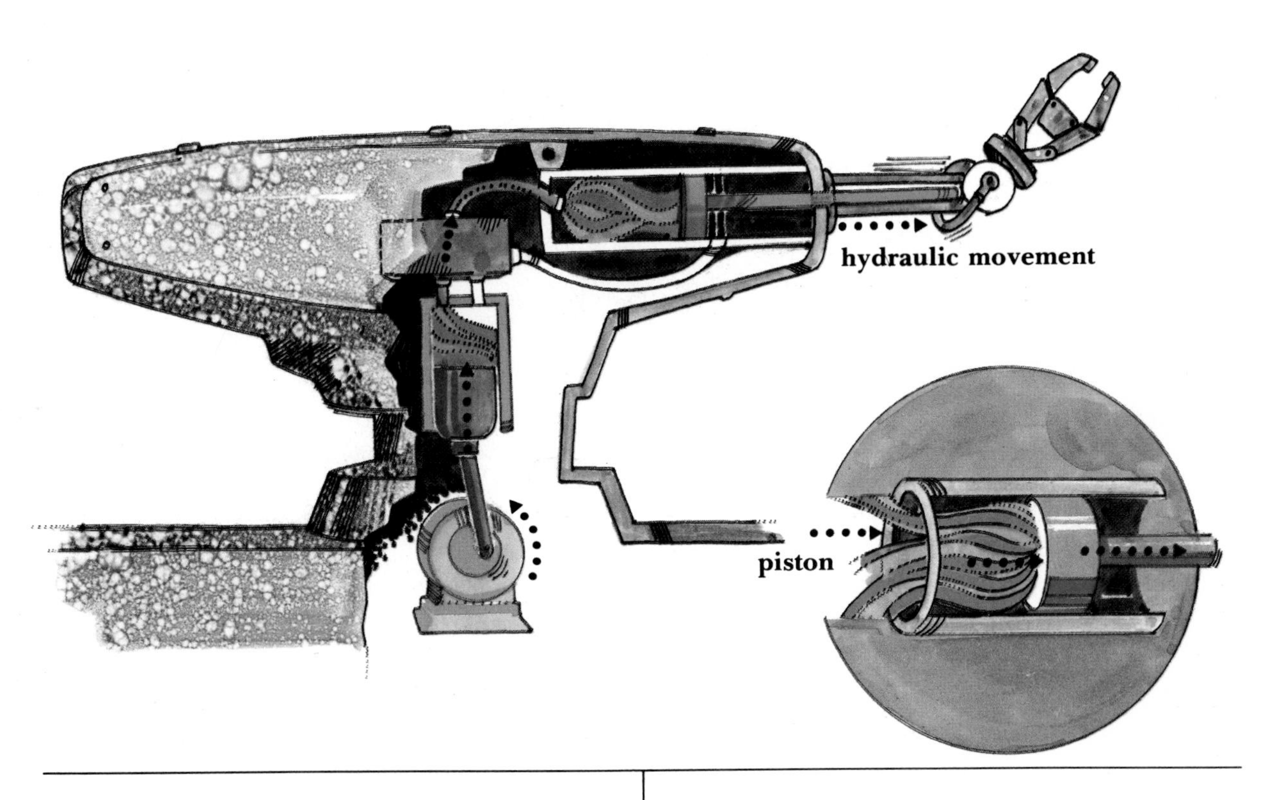

The second way the parts of a robot can move is *hydraulic*. Hydraulic means "moved with water." But the people who make robots don't use water. They use a thin kind of oil instead. The spinning motor works a pump. The pump sends the oil through hoses to small cylinders all over the robot body. Cylinders have a round body with a moving part called a piston inside them. When a cylinder fills up with oil, the piston is pushed forward. There are many cylinders in the robot.

The computer opens and shuts different valves, which allow oil to flow to certain cylinders and not to others. This makes the robot move in exactly the way the computer tells it to move.

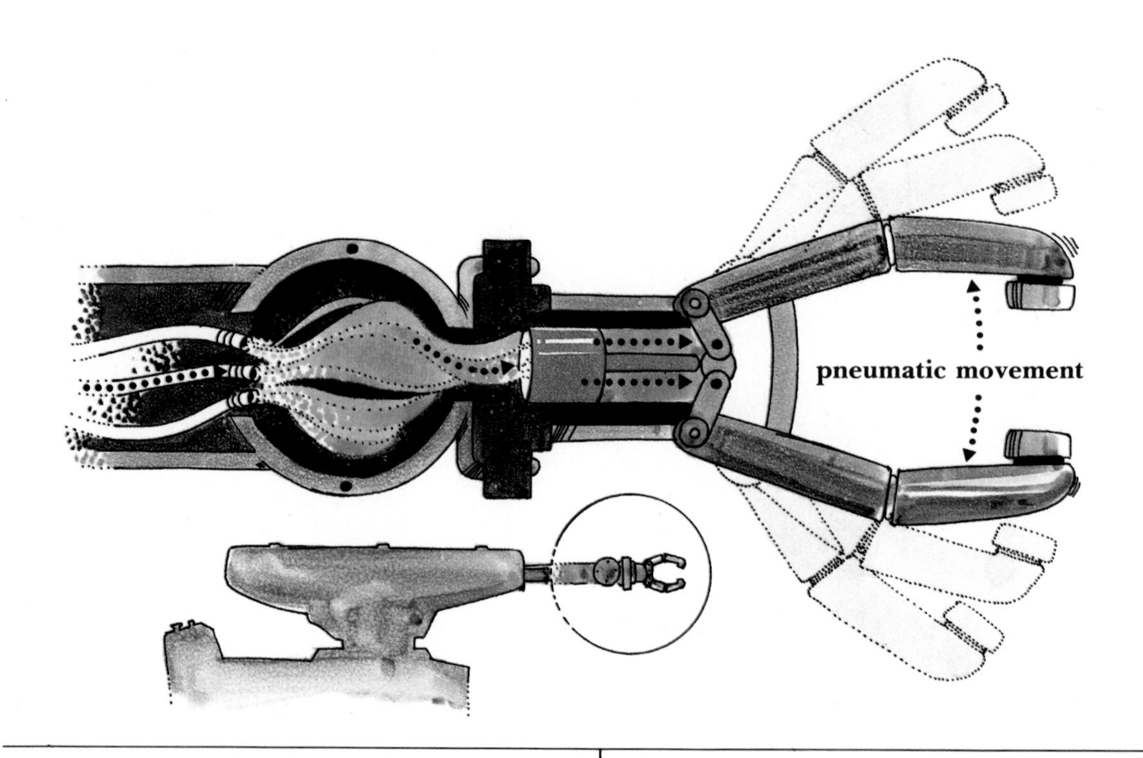

The third way robots can move is called *pneumatic*. Pneumatic means "moved with air." Pneumatic robots work just like hydraulic ones. The only difference is that they pump air into the cylinders instead of oil. The cylinders that use air are usually smaller, lighter, and less powerful than the ones that use oil.

Robots can use all three types of movements. The hydraulic cylinders are good for large, powerful movements. They are used when the robot must lift something heavy.

The mechanical and pneumatic systems are used for smaller movements that do not need as much force. The hand of a robot, for example, can be both mechanical and pneumatic. It can reach out, grip objects, and move them from one place to another.

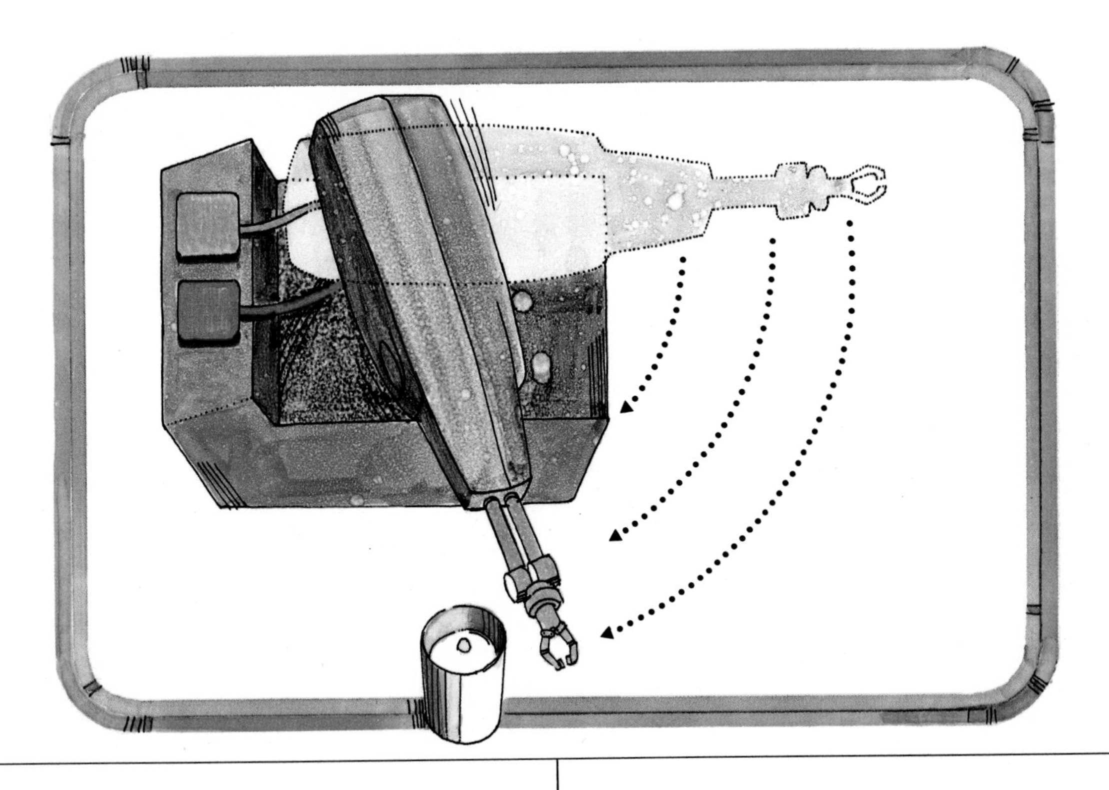

The robot has to be able to start moving. It also has to be able to stop itself. To know when to stop, the computer needs to be told how far the robot arm or hand has moved. A position sensor, an electrical device in the moving part of the robot, sends information to the computer along electrical wires. It tells the computer how far the part has moved.

The robot's computer is built into the base of the robot. Each robot has its own small computer. But a group of robots can also be hooked up to one large computer system, as if they were sharing a more powerful brain. Robots can work on the same truck body from different sides. They are controlled by the same computer so they can work with each other.

TEACHING A ROBOT

The computers used in some robots are called digital computers. They have many electrical circuits. Each circuit is a tiny maze of pathways for electricity to follow.

The computer's memory is a group of circuits with thousands of tiny magnetic switches, each with two possible positions. In one position they represent "ON" and in the other position they represent "OFF." The computer uses electricity to turn them on or off and record how they are set. Every letter, number, and act that a robot can do is translated into a long series of ONs and OFFs inside the computer. For instance, eight ONs followed by four OFFs might mean the letter "A." Ten ONs followed by two OFFs might mean the command to move the robot arm to the right.

People decide which patterns of ONs and OFFs are going to represent which letters, numbers, and actions. Instead of setting each switch inside the computer separately, people use certain commands to arrange whole bunches of switches at once. These commands are given in a programming language — a language which is easy for people to understand. However, the computer needs instructions in machine language-patterns of ONs and OFFs. One part of the computer is a machine which translates programming language into machine language. This machine is called a compiler.

Robot computers can be connected to terminals. Terminals are machines with typewriter keyboards that are used to send commands and messages to the computer so that it will know what to do.

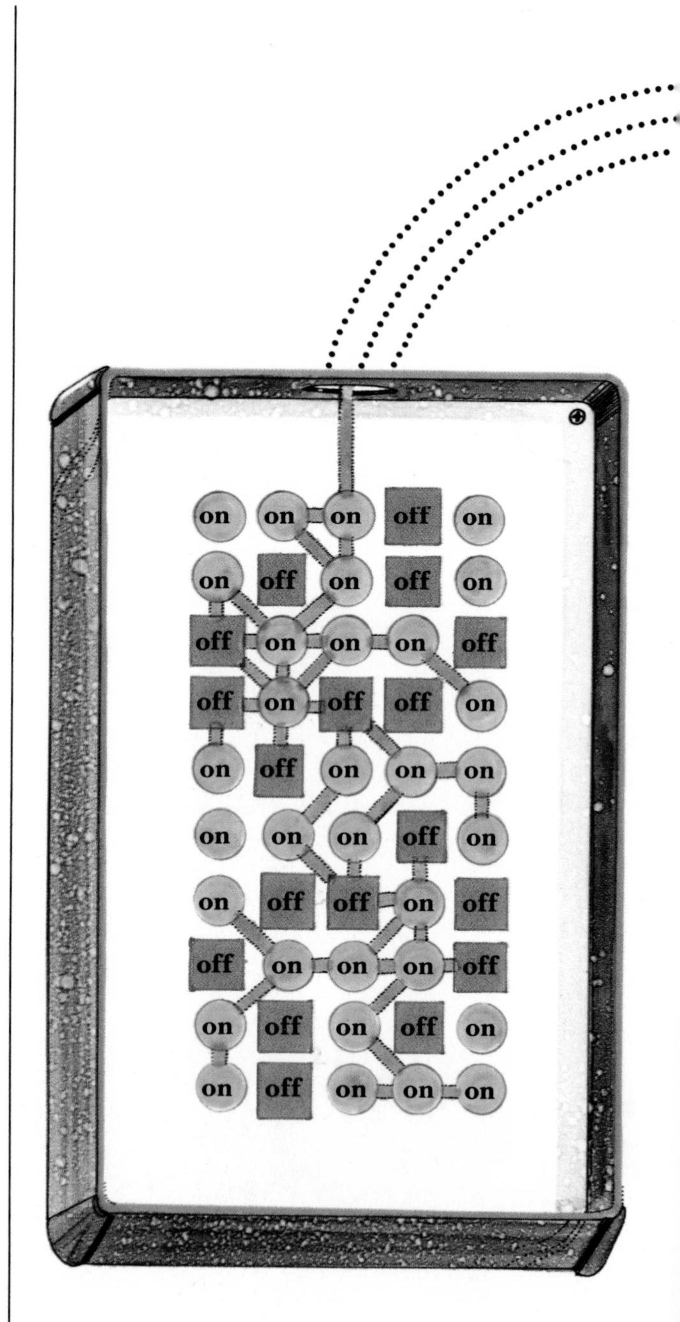

If you know a programming language, you can write a list of exact instructions to tell a computer what it should do. That is called computer programming.

With some robots, you can program a computer which guides the actions of the robot. Other robots are programmed by showing them what to do. When the robots are ready to start a job, the person who is using them shows them how to go about doing the job. This is how to show this kind of robot what to do:

Suppose you want a robot to grab a ball, lift it up, and throw it through a hoop. First you make sure that the ball and the hoop will be in exactly the right spot. That way you know the robot will be able to find them.

Then you turn on a switch on the robot's body called "teach." This switch is connected to the computer; it makes the robot ready for you to teach it. You work controls that move the robot up, down, or sideways. You move the robot through the exact set of movements it will follow later. After the robot has been moved, you turn off the teach switch, and the computer will have memorized the set of movements. From now on it will guide the robot body in following those movements.

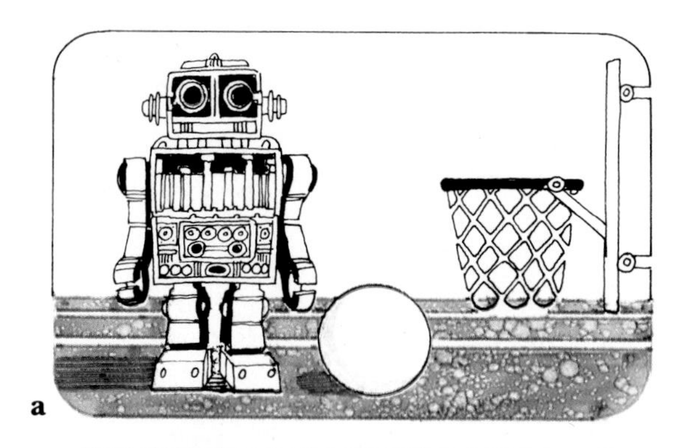

a

b

c

d

e

f

g

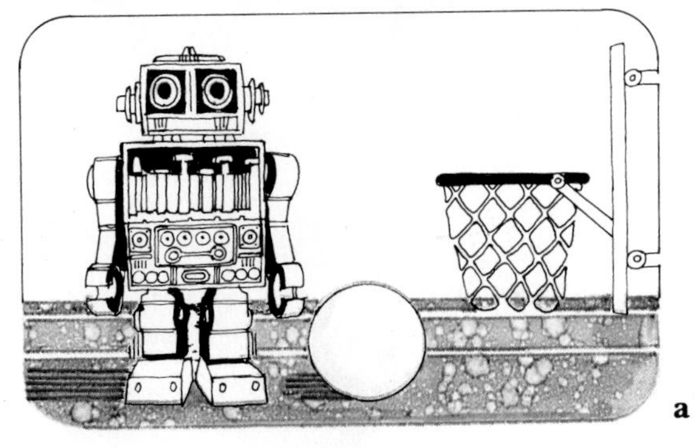

a

You bring the arm down to the ball. You make the hand open wide enough to hold the ball. You move the arm further down until the hand is around the ball. You close the grip on the ball. You move the arm with the ball, aiming toward the hoop. When it is time to let go, you open the hand. The ball flies toward the hoop.

Then you turn off the teach switch and turn on the "repeat" switch. When you start the robot, it will repeat the same movements. Every time a ball is ready to be picked up, the robot will pick it up and throw it at the hoop.

THE MIND OF A ROBOT

The most advanced computers are not attached to robots. They are bigger, more complicated sets of switching. Scientists are trying to make computers that will think like human brains. The difference will be that the computers will think with tiny switches instead of our human brain cells. Some scientists think that our brain cells work the same way that the tiny switches do in computers.

The computers that think like people will be able to do more than normal computers. They will learn from their mistakes and decide for themselves what is important to know and do. Scientists say this type of computer will have *artificial intelligence.*

Some scientists are trying to connect artificial intelligence to robots. Smarter robots will need smarter computers to do complicated things. Some things that seem easy to people would be very hard for robots to learn. For instance, understanding the many different meanings a word like "go" can have, and knowing when to use which meaning, is too complicated for robots.

Did you go to Hawaii?

The river goes from Spain to France.

What's going on?

This watch doesn't go.

Let's go fishing.

I think it's going to go well.

Will the prize go to Jeffrey?

I'll go steady with you.

Suppose you wanted to tell a friend how to do something simple, like make toast and jelly. You might tell him or her this:

"Take the jelly out of the refrigerator. Toast the bread. Smear the jelly on the bread and eat it."

But to teach a robot how to make toast and jelly you would have to tell it everything. You would have to explain where the bread is, how to open the jar and bag, and exactly how to put the bread and jelly together. Your friend knows all this because he or she has been looking and listening to other people since he

or she was born. But robots only know what a human has told them.

This is how you might tell a robot to make toast with jelly:

"Start. Go to the refrigerator. Pull the refrigerator door handle until the door opens. The jar of jelly is 15 cm tall, shaped like a cylinder, and made out of glass. It's hollow inside, and filled with purple jelly. Look at all the objects on each shelf to see if the jar of jelly is one of them. Is it on the first shelf? If yes, take it. (To take it, reach your hand forward, close your fingers tightly around it, and pull your hand back, without opening your fingers.) If no, is it on the next shelf? If yes, take it. If no, continue to the next shelf. Repeat this pattern until the last shelf. Do you have the jelly? If no, stop. If yes, hold the jelly with your fingers closed around it. Push the refrigerator door shut gently.

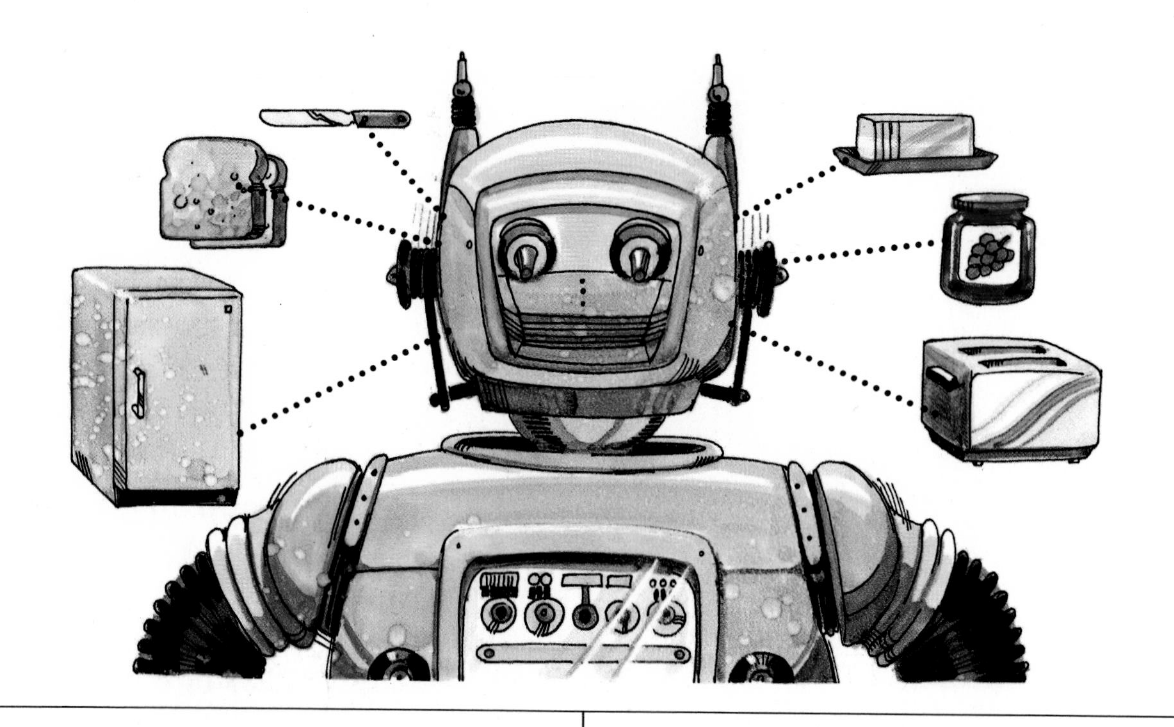

“Walk to the kitchen table with your fingers still closed around the jelly. Look for any empty flat spot on the table with an area equal to or larger than the bottom of the jar of jelly. Do you see one? If no, stop. If yes, push your hand toward that spot, until your hand is directly over that spot; then stop pushing. Then lower the jar of jelly onto the table, slowly enough so it does not break. Open your fingers until they do not touch the jar of jelly. Pull back your hand.

“Go to the pantry. The bread is about 30 cm long, 10 cm high, and 10 cm wide, inside a loose plastic bag. It should be on the third shelf from the top, against the left-hand wall. Is it on the third shelf from the top, against the left-hand wall? If no, stop. If yes, push your arm forward and close your fingers tightly around the end of the plastic bag where it is twisted together. Lift the bread up until it is off the shelf. Pull it toward you. Go to the kitchen table.”

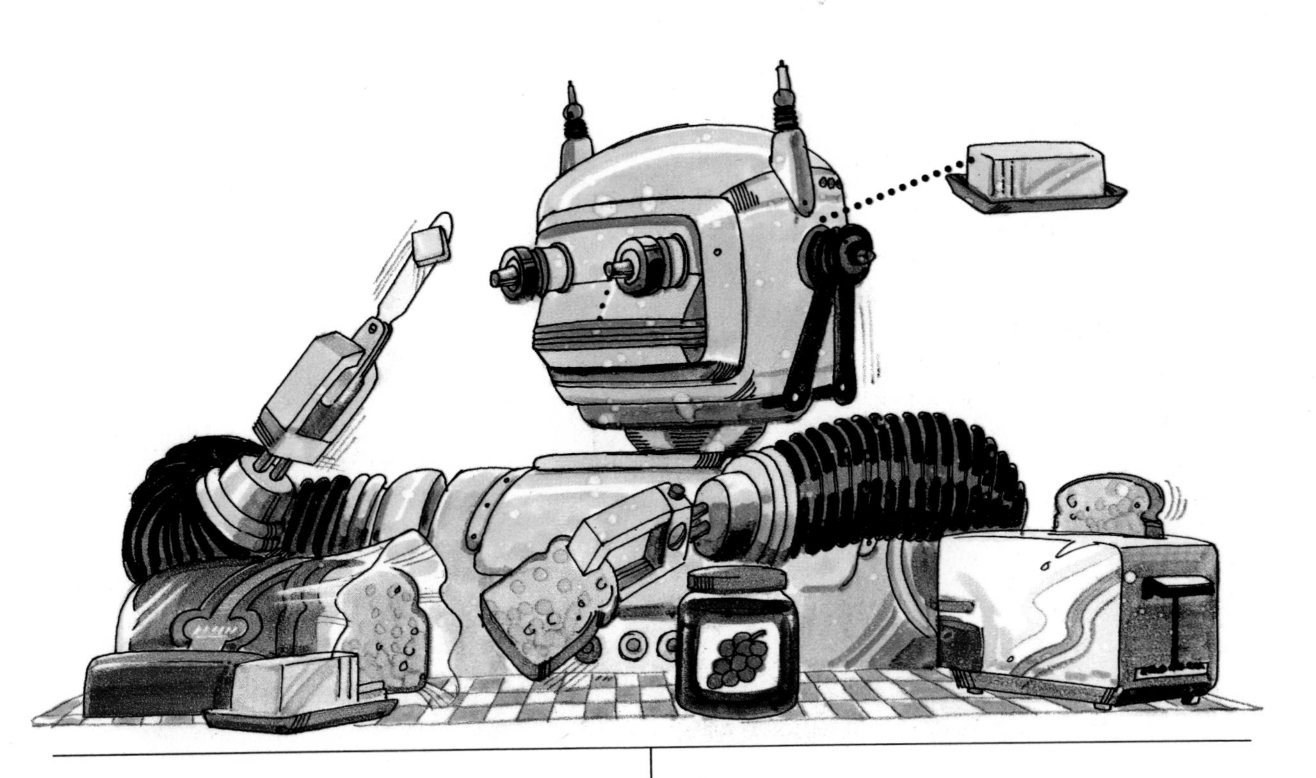

Are you bored yet? The robot has not even opened the bread, toasted it, or put jelly on it! It takes too long to describe the whole process, so we will stop here. But you probably can see by now that it's much easier to just go ahead and make the toast yourself than to tell a robot how to do it. However, you only have to teach a robot how to make toast with jelly once. Once it knows, it can make toast with jelly exactly the same way every day for years.

Most computers and robots have to be programmed very carefully, with no room for mistakes. The computer can't guess what to do if something happens that it doesn't expect. Scientists are hoping that artificial intelligence can be improved, and robots will be able to figure out more details for themselves.

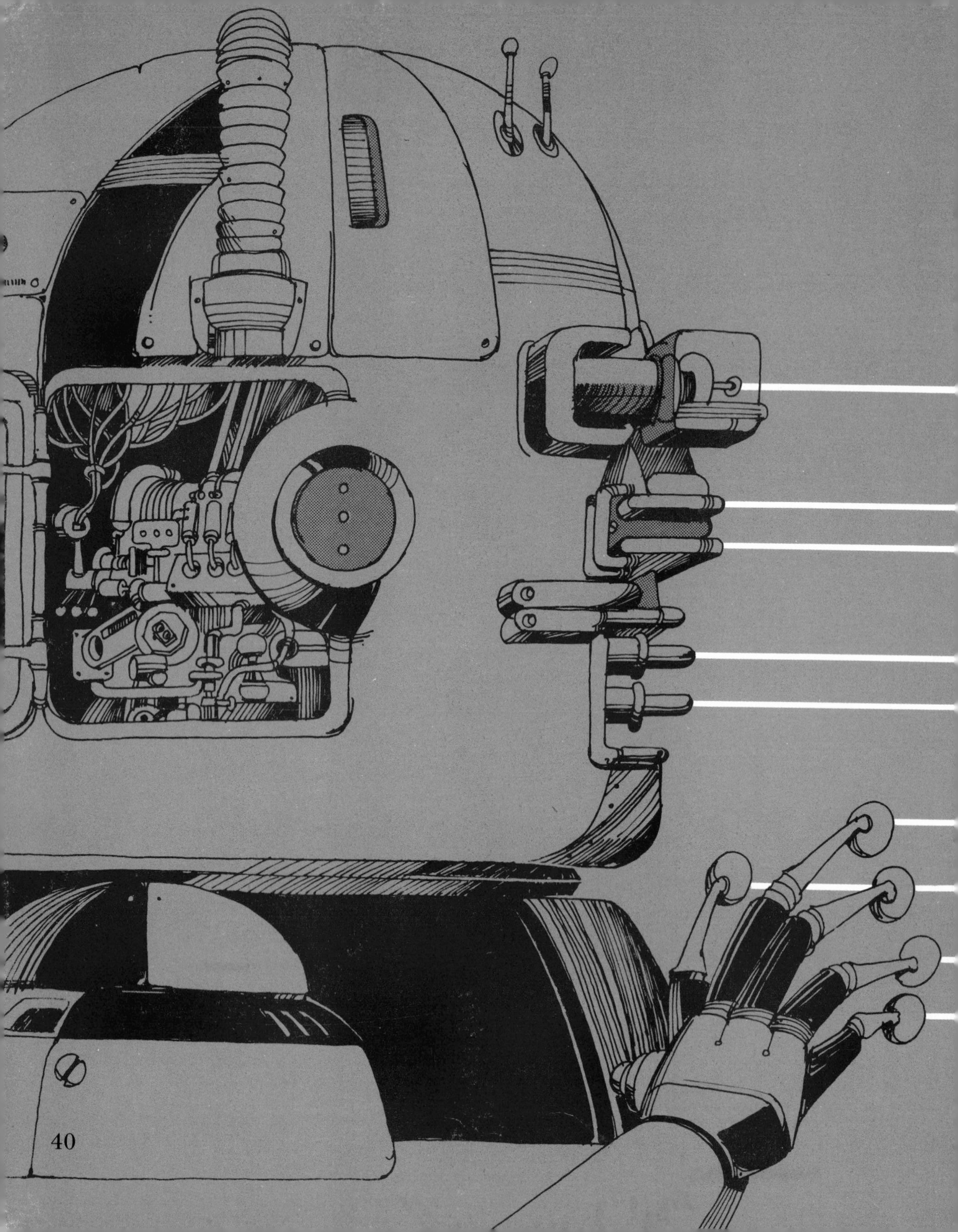

THE SENSES OF A ROBOT

Unlike people, most robots are not aware of what is going on around them. They have no senses: they can't see, hear, taste, smell, or feel what they touch. Scientists are building machines which will take the place of senses electronically. These machines are called sensors.

Robots need sensors for many reasons. They need sight sensors so they can move from place to place without bumping into things. Right now most robots that move must stay on a path. Either they follow a magnetic path in the floor, like the mail robots, or they are placed on tracks, like a railroad train.

If robots could see, they could pick up an object without needing a human or another machine to make sure it is in the right spot. For example, robots that weld metal parts cannot "see" what they are welding. The metal has to be placed exactly right for them to work on it.

If robots could hear, they could be taught to understand voices, and even to speak back to people.

Scientists have been working on sensors for robots for years. Many factory robots have simple sensors built in to measure temperature, or tell if they are holding something. But most of the sight and sound sensors are not ready to be built into actual robots yet.

A sight sensor would work like a television camera. A television camera translates a picture into a group of lines or dots. The lines or dots are so tiny that you can't see them unless you look through a magnifying glass. A computer can read each dot separately and translate the picture into its own ON-OFF language.

The hard part in making sight sensors is teaching the computer to recognize what is in a picture. We are so used to seeing objects that we recognize them if they are close to us, if they are far

from us, if they are turned sideways, and even if part of them is hidden from us. But a computer has not spent as many years as we have growing up and learning to recognize things. It is hard to teach it to recognize all these different patterns of lines as the same chair, especially when part of the chair is hidden behind a table.

There are many different types of touch sensors. Some robots have springs built into the grips or fingers on the end of their arms. If there is pressure on the springs, then the computer is informed that the robot is holding something. A thermometer can be built into a robot's arm to tell it that it is near something hot or cold.

Sometimes sensors are not built into a robot at all. They are placed near the robot's arm on a post or wall. The robot arm holds an object up to the sensor to tell if the object is hot or cold, large or small, bright or dark.

OUR FUTURE WITH ROBOTS

As robots are made that can do different things, they may take over all our factory work. Already there is a factory in Japan where only ten people are needed. Those people spend their time watching over the robots and other machines. Robot makers say that factory work is boring and hard for people to do. But it is one kind of work that people can do without going to college. Some people are afraid that only people who spend years going to school will be able to find good jobs.

Instead of working in factories, people may find other kinds of work to do. Some will work with computers. Others will make crafts by hand and sell them to people who want something different. Perhaps robots will end up doing all the work, and people will not have to work at all.

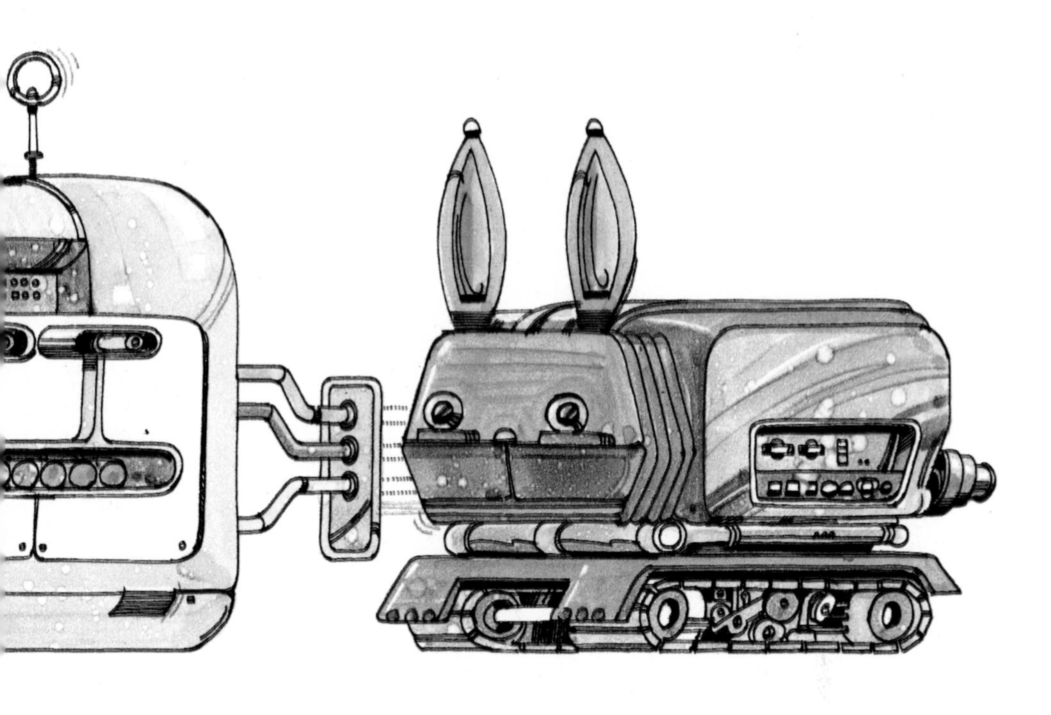

Sometimes people get scared that robots will hate working for people. Then the robots will get together, fight the humans and win, and make us work for them. That will not happen because robots are machines and have no feelings. They will always do the work they are programmed to do.

Robots may become our pets, playmates, and housekeepers. Many science-fiction stories are about housekeeping robots who become friends with their human owners. There will probably not be robots who clean houses or take care of kids for many years, because those are very complicated jobs. Besides, robots are too expensive for most people.

But robot playmates may soon run around the house and make noises, like the little tortoise. They might be able to do jobs or play games.

There will be a lot of robots very soon. Robots will probably be cheap enough for most companies in America to use in as little as ten years. They will travel to outer space and under the ocean. They may help us design entirely new kinds of other machines. They may be built into robotic cars and airplanes. And people who lose their arms or legs may have them replaced with robotic attachments.

To be ready, people should start thinking now about what we want robots to do. What do you think robots should do during the next ten years? Should they take over most kinds of work people do now? Will people like being taken care of, or will they get bored because there's nothing to do? How can robots help people have fun? What would a world full of robots be like?

PRONUNCIATION GUIDE

These symbols have the same sound as the darker letters in the sample words.

ə b**a**lloon, **a**go
a m**a**p, h**a**ve
ä f**a**ther, c**a**r
b **b**all, ri**b**
d **d**i**d**, a**dd**
e b**e**ll, g**e**t
f **f**an, so**f**t
g **g**ood, bi**g**
h **h**urt, a**h**ead
i r**i**p, **i**ll
ī s**i**de, sk**y**
j **j**oin, **g**erm
k **k**ing, as**k**
l **l**et, coo**l**
m **m**an, sa**m**e
n **n**o, tur**n**
ō c**o**ne, kn**ow**
ȯ **a**ll, s**aw**
p **p**art, scra**p**
r **r**oot, ti**r**e
s **s**o, pre**ss**
sh **sh**oot, ma**ch**ine
t **t**o, s**t**and
ü p**oo**l, l**o**se
u̇ p**u**t, b**oo**k
v **v**iew, gi**v**e
w **w**ood, glo**w**ing
y **y**es, **y**ear
ʹ strong accent
ʹ weak accent

GLOSSARY

These words are defined the way they are used in the book.

artificial intelligence (ärtʹ ə fishʹ əl in telʹ ə jəns) the ability of a computer to "think" and make decisions the way a human brain does

automaton (ȯ tämʹ ət ən) a machine made to look and act like a person or animal

circuit (sərʹ kət) a surface that has tiny pathways for electricity to follow

compiler (kəm pīlʹ ər) part of a computer that translates a programming language into a machine language of ONs and OFFs

computer (kəm pyütʹ ər) an electronic machine which can do complicated arithmetic quickly, hold the answers, and bring them out when they are needed

cylinder (silʹ ən dər) a round, hollow part with a tight-fitting piston inside that is moved by liquid or air

hydraulic (hī dröʹ lik) moved by a liquid such as water

language (langʹ gwij) the group of signs that stand for messages and commands to a computer

mechanical (mi kan′ i kəl) moved by machines such as gears and pulleys

pneumatic (nů mat′ ik) moved by air

programming (prō′ gram′ ing) the act of giving instructions to a computer

robot (rō′ bät′) a machine with moving parts, controlled by a computer, that can be taught to work by itself

robotic (rō bät′ ik) having to do with, or being similar to, robots

sensor (sen′ sȯr′) electronic machines that imitate human senses by learning about what is around them

terminal (tər′ mən əl) a machine that sends information to a computer

valve (valv) a device with a movable part that controls the flow of a liquid or gas

INDEX

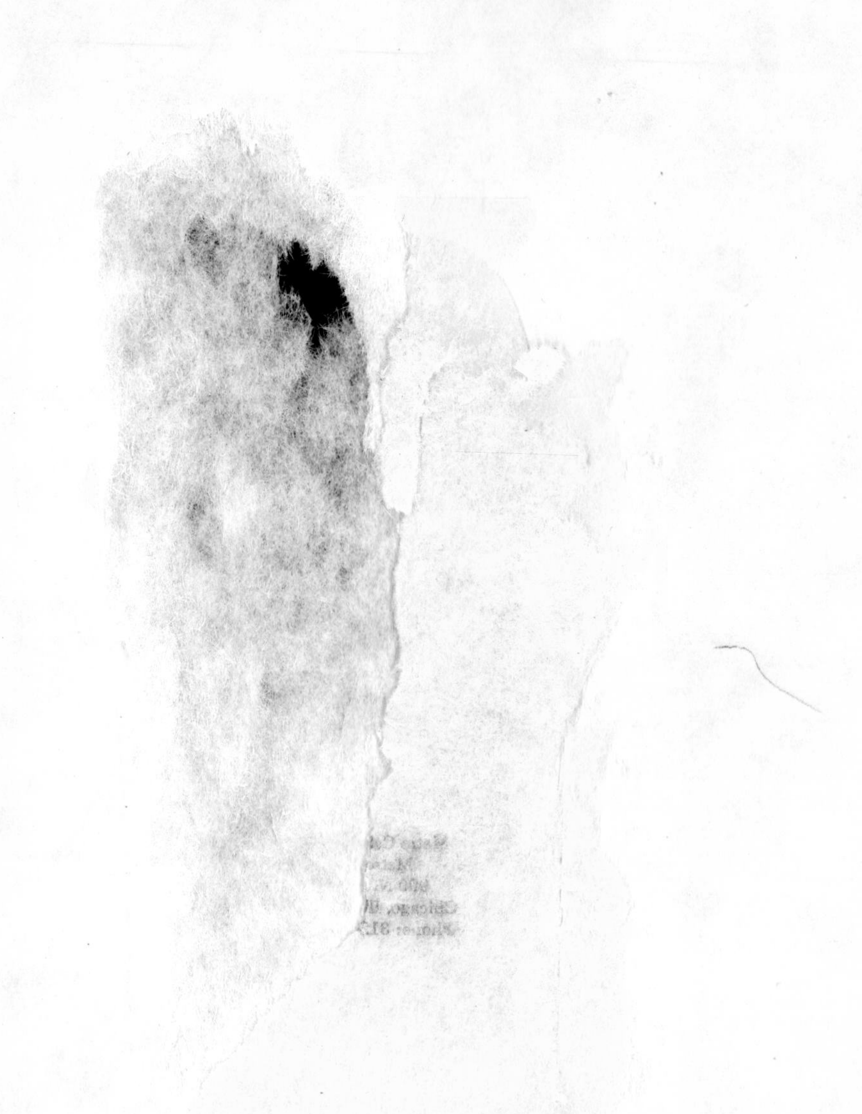